50 Flashcards $8.95

FlashCards

ARIZONA

High School

MATHEMATICS

D1136864

**Preparing Students
for the AIMS**

Printed in USA. Minimal packaging for a healthy environment.

HOLLANDAYS
Publishing Corporation

Define absolute value.

Simplify:

a) $|3|$ = _____

b) $|-3|$ = _____

c) $3|-3|$ = _____

d) $\dfrac{|-3|}{3}$ = _____

An absolute value represents a value's distance from zero.

This value is always positive.

a) 3   b) 3   c) 9   d) 1

Simplify:

1) - 4 × 3

2) - 8 ÷ 2

3) 3 × - 2

4) 25 ÷ 5

5) - 6 × - 3

6) When multiplying and/or dividing positive and negative numbers, if both numbers have the same sign, the answer is _____; however, if the two numbers have different signs, the answer is _____.

1) - 12
2) - 4
3) - 6
4) 5
5) 18
6) positive; negative

Simplify.

Which law of exponents is used in each?

1) $2^3 \cdot 2^{10} = 2^{13}$

2) $\dfrac{2^{10}}{2^3} = 2^7$

3) $(2^3)^{10} = 2^{30}$

1) 2^{13}

When multiplying like bases, add exponents.

2) 2^7

When dividing like bases, subtract exponents.

3) 2^{30}

An exponent raised to a power is multiplied by the power.

Consider these numbers:

1) 0.00456   2) 9380000

a) Explain how to convert each number to scientific notation.

b) Express each number in scientific notation.

1) a) Move the decimal point three places to the right (4.56). Since the original number was less than one and the decimal point was moved three places, the exponent would be -3.

b) Answer: 4.56×10^{-3}

2) a) Move the decimal point six places to the left (9.38). Since the original number was greater than 10 and the decimal point was moved six places, the exponent would be 6.

b) Answer: 9.38×10^{6}

Use scientific notation to simplify the following.

a) $2{,}300 \cdot 4{,}400$

b) $\dfrac{7.2 \times 10^4}{6 \times 10^{-1}}$

a) $2.3 \times 10^3 \cdot 4.4 \times 10^4 = 10.12 \times 10^7 =$
   1.012×10^8

b) 1.2×10^5

1) List four important subsets of the real number system.

2) Give a short description of each subset.

3) Give an example of each subset.

a) **Rational numbers** – numbers that can be expressed as ratios of two integers. Example: fractions

b) **Irrational numbers** – numbers that cannot be written as a ratio of two integers. Example: square root of a nonperfect square, non-repeating and non-terminating decimals

c) **Whole numbers** – the set of non-negative integers. Example: positive integers and 0

d) **Integers** – whole numbers and their opposites. Example: -5, 0, 3, 5, -3

Simplify the following using the correct order of operations.

1) $3(7-4)^2 - (6+3)$

2) $(9-3) \div 2(13-12)$

3) $4(x \cdot 3) + 3(x+2)$

1) 18

2) 3

3) 15x + 6

Evaluate each expression.

a = 3   b = 6   c = -4

1) a | -b |

2) c | -a / b |

3) | c / a |

4) | a (-b)(-c) |

1) $3 \cdot 6 = 18$

2) $-4 \cdot |-6/3| = -4 \cdot 2 = -8$

3) $|-4/3| = 4/3$

4) $|3(-6)(4)| = |-72| = 72$

Free Throw Shooting Practice - October

Player	Goal for October	Number completed	Percentage completed	Fraction completed
Byron	2000	1500		3/4
Mark	3000		50%	
Stephen	2000			1/4
Kevin	3000			1/3
Chris	3000	2000		

Create a bar graph that shows free throws completed.

Byron	2000	1500	75%	3/4
Mark	3000	1500	50%	1/2
Stephen	2000	500	25%	1/4
Kevin	3000	1000	33.3̄%	1/3
Chris	3000	2000	66.6̄%	2/3

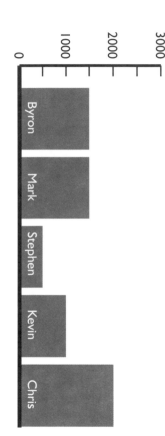

Suppose you write all the letters of the word **protractor** on individual cards and place them in a bag. If you take a card from the bag without looking, what is the probability of picking a card with the following letters?

a) R   b) O   c) P   d) S

a) 3/10

b) 2/10 or 1/5

c) 1/10

d) 0

The graph shows the sales trend for the first three weeks of a local band's album release.

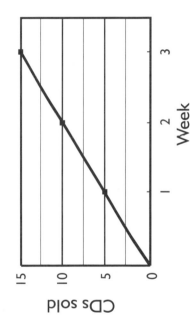

How are the sales changing? At what rate?

The sales are increasing at a rate of about 5 CDs a week.

What is the difference between a **combination** and a **permutation**?

a) Greg tosses a two-sided coin twice. How many **combinations** of heads and tails can he get as a result?

b) How many **permutations** can he get?

c) Express the possible outcomes in a tree diagram.

Combination: order of terms does not matter

Permutation: order of terms matters

a) Greg can have 3 different combinations: H,H T,T and H,T *or* T,H (T,H and H,T are considered the same combination because order does not matter)

b) Greg can get 4 different permutations: H,H T,T H,T *and* T,H (H,T and T,H are considered two different permutations because order does matter)

c)

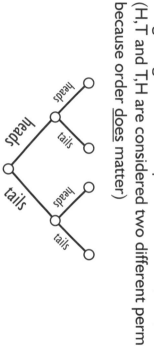

1) Define and describe the "line of best fit" in a scatterplot.

2) How do you find the slope of the line of best fit in a scatterplot?

3) What kind of correlation is illustrated in each data set below?

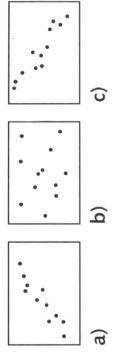

a)

b)

c)

1) A "line of best fit" is a trend line that most closely represents the data on a scatterplot.

2) Pick two points on the line of best fit. Determine the slope by finding the value of the rise ÷ run.

3) a) Positive correlation
   b) No correlation
   c) Negative correlation

Basic Counting Principle

1) Define the Basic Counting Principle.

2) When is it used?

Sample Space

Name four different ways to gather a sample.

Basic Counting Principle

1) Use this principle to find all possible outcomes by multiplying the number of ways each event can occur.

2) It is used to determine the total number of outcomes in a sample space.

Sample Space

(1) Random sample

(2) Survey response

(3) Convenience sample

(4) Representative sample

Define:

1) mean

2) median

3) mode

4) range

5) The first three terms are referred to as _____.

6) Which measure of central tendency is affected most by extremes in data?

1) The **mean** is the arithmetic average of a set of numbers.

2) After arranging the data values in order, the **median** is the middle value.

3) The **mode** is the value or values that occur most often in a data set.

4) The **range** is the distance between the highest and lowest values in a data set.

5) measures of central tendency

6) mean

1) Represent the data in a stem and leaf graph.

The following grades were scored on Mr. Saluke's Algebra I test:

56, 96, 100, 98, 93, 82, 80, 75, 97, 68, 73, 88, 97, 82, 73, 91, 97, 86, 90, 85

2) Find the mean, median, mode, and range.

1)

stem	leaf
5	6
6	8
7	3 3 5
8	0 2 2 5 6 8
9	0 1 3 6 7 7 7 8
10	0

2) Mean = 85.35
Median = 87
Mode = 97
Range = 56-100

Find the perimeter for the following figure.

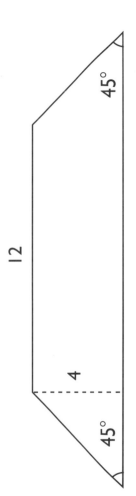

12

4

45°

45°

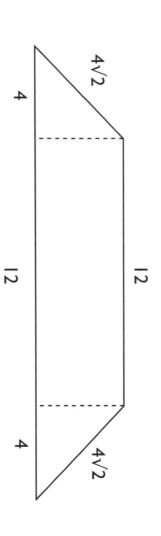

Perimeter = 32 + 8√2

On the graph of a linear inequality, how do you decide:

a) whether the line is solid or dashed

b) where to shade the graph

c) study the graph, then complete this inequality:

$y \underline{\quad} x + 1$

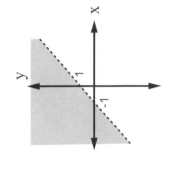

a) If the linear inequality uses a $\geq$ or $\leq$, the graph is a solid line.
If the linear inequality uses a $<$ or $>$, the graph is a dashed line.

b) Pick a point on either side of the graphed line of the inequality.
Substitute the x and y coordinates of that point into the inequality.

• If these substituted values yield a true statement, shade the graph on the side of the graph where the point is located.

• If the statement is false, shade the side of the graph opposite the point.

c) $y > x + 1$

A local movie theatre charges $9.50 for adults and $7.00 for children.

1) Write an expression that represents the total cost for a group of **A** adults and **C** children to go to the movies.

2) How would it cost for a husband and wife to take their three children to a movie?

1) 9.5A + 7C

2) 9.5 (2) + 7(3) = $40

Write each trigonometric ratio
using the sides of a right triangle.

1) Sin A = ____

2) Cos A = ____

3) Tan A = ____

1) SOH → sin A = $\frac{\text{opposite}}{\text{hypotenuse}}$ = 5/13

2) CAH → cos A = $\frac{\text{adjacent}}{\text{hypotenuse}}$ = 12/13

3) TOA → tan A = $\frac{\text{opposite}}{\text{adjacent}}$ = 5/12

Solve:

1) $x^2 = 36$

2) $(x-1)^2 = 25$

3) $x^2 + 6x = -9$

4) $2x^2 + x - 3 = 0$

1) $x = \pm 6$

2) $x = 1 \pm 5 = 6, -4$

3) $x = -3$

4) $x = -3/2, 1$

Given the following sequences, find the common difference or the common ratio and predict the 10th term.

a) 1, 4, 7, 10...

b) 2, 6, 18, 54...

a) common difference = 3
10th term = 28
(1, 4, 7, 10, 13, 16, 19, 22, 25, 28)

b) common ratio = 3
10th term = 39,366
(2, 6, 18, 54, 162, 486, 1458, 4374, 13122, 39366)

Simplify.

1) $\begin{bmatrix} 3 & 4 \\ 1 & -1 \end{bmatrix} + \begin{bmatrix} 6 & 0 \\ 9 & 4 \end{bmatrix} =$

2) $\begin{bmatrix} 3 & 4 \\ 1 & -1 \end{bmatrix} - \begin{bmatrix} 6 & 0 \\ 9 & 4 \end{bmatrix} =$

3) $5\begin{bmatrix} 2 & 1 & 7 \\ -3 & 0 & 4 \end{bmatrix}$

1) $\begin{bmatrix} 9 & 4 \\ 10 & 3 \end{bmatrix}$

2) $\begin{bmatrix} -3 & 4 \\ -8 & 5 \end{bmatrix}$

3) $\begin{bmatrix} 10 & 5 & 35 \\ -15 & 0 & 20 \end{bmatrix}$

Solve the system of equations using a matrix:

$x + y = 10$

$x - y = 6$

$$\begin{bmatrix} 1 & 1 & | & 10 \\ 1 & -1 & | & 6 \end{bmatrix} \overset{①}{=} \begin{bmatrix} 2 & 0 & 16 \\ 1 & -1 & 6 \end{bmatrix} \overset{②}{=} \begin{bmatrix} 1 & 0 & 8 \\ 1 & -1 & 6 \end{bmatrix}$$

$$\overset{③}{=} \begin{bmatrix} 1 & 0 & 8 \\ 0 & -1 & -2 \end{bmatrix} \overset{④}{=} \begin{bmatrix} 1 & 0 & | & 8 \\ 0 & 1 & | & -2 \end{bmatrix} \quad \begin{array}{l} x = 8 \\ y = 2 \end{array}$$

① Add bottom row to top.

② Divide top row by 2.

③ Multiply top row by -1. Add top row to bottom.

④ Multiply bottom row by -1

Simplify.

1) $8x - 3y + 6 \cdot x \div 3 \cdot 4$

2) $\dfrac{(3x^2) - y^2 - 3x}{x}$

1) 16x − 3y

2) -y²

1) The problems below are examples of
_____. What rule must one
remember when working with this type
of problem?

2) Solve

a) $|x - 5| \leq -7$

b) $-\frac{2}{3}x < 4$

c) $-3x > 9$

d) $x + 3 \geq -8$

1) inequalities; When multiplying or dividing by a negative number, switch the inequality sign.

2) a) $-2 \leq x \leq 12$
b) $x > -6$
c) $x < -3$
d) $x \geq -11$

Given the function, state the domain and the range.

a)

x	y
0	1
-3	-5
2	5

b) $f(x) = x - 1$

c) $f(x) = x^2 + 1$

a) **Domain:** 0, -3, 2 **Range:** 1, -5, 5

b) **Domain:** any real number; $\{x : x$ is any real number$\}$

   Range: any real number; $\{y : y$ is any real number$\}$

c) **Domain:** any real number; $\{x : x$ is any real number$\}$

   Range: positive numbers greater than or equal to one; $\{y : y \geq 1\}$

1) What do parallel lines have in common?

2) Given the following pairs of lines, determine which are parallel.

a) $x + 3y = 8$
   $12y = -4x + 32$

b) $x + y = -4$
   $x + y = 2$

c) $12x - 4y = 1$
   $6x - 3y = 2$

1) Parallel lines have the same slope.

2) a) parallel
   b) parallel
   c) not parallel

Simplify.

a) $\sqrt{25}$ b) $\sqrt{200}$ c) $\sqrt{72x^2y^3z^5}$

Find the missing values in the proportions:

d) $\dfrac{3}{7} = \dfrac{x}{21}$ e) $\dfrac{2}{3} = \dfrac{x}{27}$ f) $\dfrac{4}{9} = \dfrac{12}{x}$

a) 5

b) $10\sqrt{2}$

c) $6xyz^2\sqrt{2yz}$

d) 9

e) 18

f) 27

1) Explain the process used to convert a linear equation to slope-intercept form.

2) Tell the relationship of the slopes of two lines if they are:

   a) parallel

   b) perpendicular

1) Solve the equation for y and arrange the equation in slope-intercept form, $y = mx + b$.

2) a) Slopes are equal.

b) Slopes are negative reciprocals of each other. Example: If the slope of one line is $\frac{2}{3}$, the slope of a line perpendicular to it is $-\frac{3}{2}$.

Solve:

a) $3(x - 2) + 1 = 2x - 5$

b) $3x + 5x - 10 = 10 - 12x$

c) $3x - 2x + 5 = 15 - 3$

d) $x + 9x - 6x = 20 - 4$

a) $3(x-2)+1 = 2x-5$
$3x-6+1 = 2x-5$
$3x-5+5 = 2x-5+5$
$3x-2x = 0$
$x = 0$

b) $3x+5x-10 = 10-12x$
$8x-10 = 10-12x$
$8x-10+10 = 10-12x+10$
$8x+12x = 20-12x+12x$
$20x = 20$
$x = 1$

c) $3x-2x+5 = 15-3$
$x+5 = 12$
$x+5-5 = 12-5$
$x = 7$

d) $x+9x-6x = 20-4$
$10x-6x = 16$
$4x = 16$
$x = 4$

Simplify using the laws of exponents.

a) $(2xy)^0$

b) $\left(\dfrac{4x^{10}}{6y}\right)^2$

c) $(12\,a^3z^{11})(5\,a^2z^{-10}y^4)$

d) $\dfrac{15\,w^{20}\,x^{10}\,y^5}{25\,w^{25}\,x^5\,y^6}$

e) $(3\,x^2y)^2 + (7\,x\,z^3)^2$

a) 1

b) $\dfrac{4x^{20}}{9y^2}$

c) $60a^5zy^4$

d) $\dfrac{3x^5}{5w^5y}$

e) $9x^4y^2 + 49x^2z^6$

Emily's wages for a 15 hour work week increased by $6.75. Before the raise, she was earning $7.80 per hour. What is her new hourly wage?

$15x - 15(7.8) = 6.75$

$x = 8.25$

$8.25 / hour

Rewrite each sentence as a mathematical expression. Let x = weekly pay.

1) I can earn as much as $60.00 this week.

2) I am working full time next week, so I will earn more than $60.00.

1) $x \leq \$60.00$

2) $x > \$60.00$

Mr. Houston drove 60 miles in 1 1/4 hours. Answer the following questions about Mr. Houston's trip.

a) What formula can you use to find Mr. Houston's speed?

b) Mr. Houston's average speed was _____ miles per hour.

c) At the same speed, how far will Mr. Houston travel in three hours?

a) Distance = rate × time

b) Distance = rate × time

60 = rate × 1.25

60 ÷ 1.25 = rate

rate = 48 miles per hour

c) 48 × 3 = 144 miles

1) Set up the equation to solve for x. Use the Pythagorean Theorem: $a^2 + b^2 = c^2$

2) Solve for x.

36

©2008 Hollandays Publishing Corporation

1) $8^2 + 6^2 = x^2$

2) $64 + 36 = x^2$

$100 = x^2$

$10 = x$

1) How do you find the sum of the interior angles of a regular polygon?

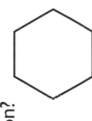

2) How do you find the sum of the exterior angles of a regular polygon?

1) First, find how many triangles can be formed by drawing diagonals from one point in the polygon to the remaining vertices.

Multiply the number of triangles (4) formed by 180° giving the sum of the measures of the interior angles of the polygon. (720°)

2) The sum of the exterior angles of a regular polygon is always 360°.

Explain what happens to a figure when the following transformations take place:

1) Rotation

2) Reflection

3) Translation

4) Dilation

1) Each point of the figure is moved through the same angle. (turn) Example: 90° rotation: $(x, y) \rightarrow (-y, x)$

2) Each point of the figure is moved through a mirror image. (flip) Example: over x-axis: $(x, y) \rightarrow (x, -y)$

3) Every point of the image is moved the same distance in the same direction. (slide) Example: $(x, y) \rightarrow (x + a, y + b)$

4) The image keeps the same shape but changes size. (enlarge/reduce) Example: $(x, y) \rightarrow (kx, ky)$

1) Find the volume of the following rectangular solids.

2) Find the surface area of the solids.

a)

8" 6" 12"

b)

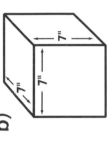

7" 7"

1a) 576 cubic inches

b) 343 cubic inches

2a) 432 square inches

b) 294 square inches

Beth is going to paint the floor of her L-shaped swimming pool.

How many square feet of paint will Beth apply?

What is the volume of her pool if it averages 5 ft deep?

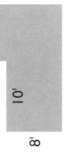

8'

8'

10'

8'

Beth will need paint to cover 208 ft^2.

The volume of the pool is 1040 ft^3.

1) When two triangles are similar, corresponding angles are _____ and corresponding sides are _____ .

2) When two triangles are congruent, corresponding angles are _____ and corresponding sides are _____ .

1) When two triangles are similar, corresponding angles are **equal** and corresponding sides are **in proportion**.

2) When two triangles are congruent, corresponding angles are **equal** and corresponding sides are **equal in length**.

Given: Circle O

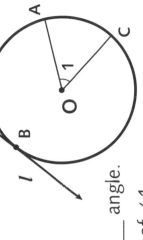

1) ∠1 is called a(n) _____ angle.

2) Describe the major arc of ∠1.

3) Describe the minor arc of ∠1.

4) How do you find the m∠1?

5) Line *l* is called a(n) _____ .

1) central

2) The major arc is from A to C around the outside of the central angle. Major arc: $\overset{\frown}{ABC}$

3) The minor arc is from A to C through the interior of the central angle. Minor arc: $\overset{\frown}{AC}$

4) The measure of a central angle is the same as the measure of its intercepted arc.
m $\angle 1$ = degree measure of $\overset{\frown}{AC}$

5) tangent

Similar Triangles
△ ABC ~ △ XYZ

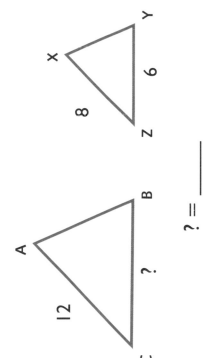

? = _____

43

Similar triangles have sides that are in proportion to each other.

Cross multiply to solve.

$$\frac{8}{12} = \frac{6}{?}$$

? = 9

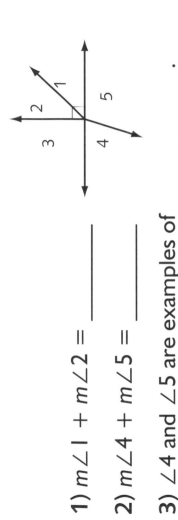

1) $m\angle 1 + m\angle 2 =$ _____

2) $m\angle 4 + m\angle 5 =$ _____

3) $\angle 4$ and $\angle 5$ are examples of _____.

4) $m\angle$__ $+ m\angle$__ $+ m\angle$__ $= 180°$

5) $\angle 1$, $\angle 2$ and $\angle 4$ are _____ angles.

1) 90°

2) 180°

3) supplementary angles

4) $m\angle 1 + m\angle 2 + m\angle 3 = 180°$

5) acute

1) Using the distance formula, find the length of the segment between points A and B.

Point A: (-7, -4)

Point B: (9, 3)

2) Given points A and B, find the coordinates of the midpoint of $\overline{AB}$.

1) $\sqrt{(-7-9)^2+(-4-3)^2}$

$-16^2+-7^2=305$

$\sqrt{305} \approx 17.46$ units

2) $\left(\dfrac{-7+9}{2}, \dfrac{-4+3}{2}\right)$

$\left(1, -\dfrac{1}{2}\right)$

Graph the following using the slope-intercept method.

1) $y = 4/5x + 1$

2) $2x + y = 3$

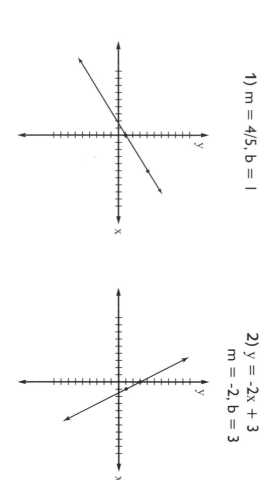

1) m = 4/5, b = 1

2) y = -2x + 3
m = -2, b = 3

Find the missing values.

47

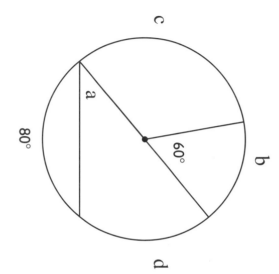

c

a

60°

b

80°

d

a = 50°

b = 60°

c = 120°

d = 100°

Find the area of the shaded portion of the circle.

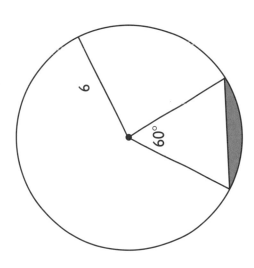

Area of the circle $= \pi(6)^2 = 36\pi = 6\pi$

$$\frac{36\pi}{6}$$

Area of the triangle $= \dfrac{6 \bullet 3\sqrt{3}}{2} = 9\sqrt{3}$

Area of the shaded region $= 6\pi - 9\sqrt{3}$

All dogs are animals.

a) write as an if-then statement

b) write the converse

c) write the contra positive

d) state whether or not each new statement is true or false

a) If I have a dog, then it is an animal. — True

b) If I have an animal, then it is a dog. — False

c) If I do not have an animal, then I do not have a dog. - True

1) Find X
2) Find Y

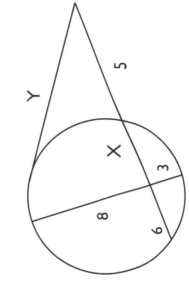

1) $X^2 = 5(6+4+5)$
$X^2 = 75$
$X = 5\sqrt{3}$

2) $8 \cdot 3 = 6Y$
$Y = 4$

ARIZONA HIGH SCHOOL QUESTION OF THE DAY

Start each class with a question that matches the **Arizona High School Exam**:

- Generate discussion of key concepts with targeted and in-depth questions

- Written by experienced teachers

- Labeled according to the standard being addressed

- Save time, improve classroom management and keep instruction and review flowing

- On transparencies in a 3-ring binder or on CD-ROM

ARIZONA ORDER FORM

High School Question of the Day

- ☐ Mathematics.. $89
- ☐ English-Language Arts............................$89
- ☐ Science.. $89
- ☐ Business.. $89
- ☐ Family & Consumer Science................$89
  Career technical teachers have unique opportunities to support academic learning in an applied context. Each interdisciplinary question is labeled according to both the career technical and academic standard being addressed.

- ☐ Site License on CD-ROM (per title)....$295
  The site license includes rights to copy the CD-ROM and project/ print from the CD-ROM.

High School Flashcards

- ☐ English-Language Arts............................$8.95
- ☐ Mathematics... $8.95
- ☐ 25 or More Sets, Each.............................$6.00

Please add 8% for shipping and handling.
Phone: 800-792-3537   Fax: 937-222-2665
Order on the web @ www.hollandays.net